This Notebook Belongs to :

AF886221

Standard Algorithms: Add, Subtract, Multiply, and Divide

Addition

1) Line up numbers by place value.
2) Add zeros as place holders if necessary. (See Example A, Step 2.)
3) Add columns from smallest to largest place value (right to left). If the sum of two digits is 10 to 19, carry the "1" in the double digit to the next place value, and add. (See Example B below.)

Example A:

Step 1)	Step 2)	Step 3)
2.351 + 1.3	2.351 + 1.300	2.351 + 1.300 3.651

Example B:

Step 1)	Step 2)	Step 3)
4,273 + 283	4,273 + 283	1 4,273 + 283 4,556

Subtraction

1) Line up numbers by place value.
2) Add zeros as place holders if necessary.
3) Subtract in columns from smallest to largest place value (right to left). If the top number is smaller than the bottom number, regroup or borrow from the column to the left (next largest place value). Repeat this process as needed.

Example A:

Step 1)	Step 2)	Step 3)
26.23 − 5.4	26.23 − 5.40	5 12 2~~6~~.~~2~~3 − 5.40 20.83

Example B:

Step 1)	Step 2)	Step 3)	
833 − 588	833 − 588	2 13 8~~3~~~~3~~ − 588 5	12 7 ~~2~~ 13 ~~8~~~~3~~~~3~~ − 588 245

Multiplication

1) Line up the numbers by place value. The larger number is on top.
2) Multiply the bottom digit in the ones place to the top digit in the ones place. From the product, write down the ones digit and carry the tens digit, if necessary to the next place value on top.
3) Multiply the bottom digit in the ones place to the top digit in the tens place and add any value that was carried. Continue multiplying the ones digit from the bottom number to the digits in the top number, right to left.
4) If the bottom number is two or more digits, repeat the process by adding a 0 to the partial product on the next line for each place value you multiply throughout. (See Example B, Step 4 below.)
5) Add the partial products resulting from multiplying by each digit in the bottom number.

Example A:

Step 1)	Step 2)	Step 3)
225 Top number × 3 Bottom number	1 225 × 3 5	1 225 × 3 675

Example B:

Step 1)	Step 2)	Step 3)
161 × 12	161 × 12 2	1 161 × 12 322

Step 4)	Step 5)
161 × 312 322 1610 Added zero	161 × 12 322 Partial product + 1610 Partial product 1932

Division

Dividend: the number being divided

Divisor: the number which will divide the dividend

Quotient: the answer to a division problem

1) Take the first digit of the dividend. Check if this digit is greater than or equal to the divisor.
2) Divide: Divide the dividend by the divisor and write the answer on top as the quotient.
3) Multiply: Multiply the answer on top by the divisor.
4) Subtract: Subtract the result from the digit and write the difference below.
5) Bring it down: Bring down the next number (if present).
6) Repeat the same process.

Step 1) The first digit of the dividend (435) is 4 and it is equal to the divisor (4).

Step 2) **Divide**: $4 \div 4 = 1$. Write 1 up top.

Step 3) **Multiply**: $1 \times 4 = 4$

Step 4) **Subtract**: $4 - 4 = 0$.

Step 5) **Bring it down**: Bring down 3, the second digit.

Step 6) Repeat the process.

Step 1) Since 3 < 4, write 0 as the quotient above and bring down the next digit in the dividend (5).

Step 2) **Divide**: $35 \div 4$ is approximately 8.

Step 3) **Multiply**: $8 \times 4 = 32$

Step 4) **Subtract**: $35 - 32 = 3$

The quotient is 108 and the remainder is 3. 108 R 3

Example A:

```
   1 0 8
4) 4 3 5
 − 4
   0 3
 −   0
     3 5
   − 3 2
       3
```

Made in the USA
Monee, IL
05 November 2026

67873794R10079